Make: Like the Pioneers

The Editors of Make:

MAKER MEDIA
SAN FRANCISCO, CA

Make: Like The Pioneers

by The Editors of Make:

Copyright © 2015 Maker Media. All rights reserved.

Printed in the United States of America.

Published by Maker Media, Inc., 1160 Battery Street East, Suite 125, San Francisco, CA 94111.

Maker Media books may be purchased for educational, business, or sales promotional use. Online editions are also available for most titles (*http://safaribooksonline.com*). For more information, contact our corporate/institutional sales department: 800-998-9938 or *corporate@oreilly.com*.

Editor: Patrick Di Justo	**Interior Designer:** David Futato
Production Editor: Nicholas Adams	**Cover Designer:** Brian Jepson
Proofreader: Amanda Kersey	**Illustrator:** Rebecca Demarest
Indexer: Bill Morrison and Megh Jones	

October 2015:   First Edition

Revision History for the First Edition

2015-09-24:   First Release

See *http://oreilly.com/catalog/errata.csp?isbn=9781680450545* for release details.

978-1-680-45054-5

[LSI]

Table of Contents

PART II. Afternoon

Preface

The book you are holding in your hand is a departure from the usual *Make:* book.

In this book you won't find any robots, drones, Arduinos, Raspberry Pi's, or smart entities in the Internet of Things.

What you will find are fire bows and homemade cider, pressed paper and hand-sewn books, crafted chairs to sit in while you read them, and a sculpted lamp to light the night.

You're holding a book about living like a pioneer—perhaps like the pioneers of the American West prior to 1890, perhaps like anyone trying to pioneer a new life for themselves with minimal assistance (or interference) from the greater civilization around us.

This book is a collection of old projects that have appeared in the pages of *Make:* over the past 10 years. But they're also projects that could have been undertaken before the 20th century. We'll take on these projects in a modern way, with contemporary tools and respect for safety.

The book begins, as a pioneer's day might, with the lighting of the fire. While many pioneers crossing the American plains would have banked their fire overnight and revived it in the morning, occasionally they would need to start a fire from scratch. Wendy Tremayne, the author of the project, shows you exactly how you can do that without a match.

The next step in the morning routine is given to a curious hybrid of bacon and soap. It's a pretty common thing to wash one's self first thing in the morning, and it's also a venerated American custom to eat bacon for breakfast. The author Tim King takes care of both traditions in one project by showing you how to make homemade soap using rendered pig fat (which you can pretend you slaughtered yourself on the Oregon Trail). Soap making was a respected profession in the colonial and pioneer eras, with the village soap maker often doubling as the local candle maker (called a chandler), because both items were created from the rendered fat of livestock, such as pigs and cows.

But you can't break your fast with bacon soap. So how about a nice glass of homemade apple cider? In this book, Dr. Nevin Stewart, a former chemist who has trained as a chef—chemistry and cooking have a lot in common, especially if you're hungry—shows you how to use modern techniques to make old-fashioned cider.

The back-to-back projects of paper making and bookbinding were lucrative professions for nearly 400 years, from the 16th through the 19th centuries. An early paper maker would have made into paper leftover scraps of cloth (usually the underwear of recently deceased people) as well as other fibrous plant scraps. In this book, François Vigneault shows you how to make paper just like they did, but without the used undies. He originally developed this *Make:* project as a way to create something beautiful and functional from junk mail!

The act of typesetting a publication—using tiny pieces of lead alloy with punched letters on them to create blocks of text—has become much less common in the 21st century. But the art of bookbinding endures: sewing together various sheets of paper into "signatures," and then securing several signatures of into a hardcover book. While the binding process for a book like this one is automated, Brian Sawyer (who calls himself a bibliophile at heart, a book editor by trade, and a book crafter by avocation) shows you how to do it by hand.

Part of a pioneer's day might have been given over to longer-term projects like crafting furniture, such as the furniture how-tos in this book by authors Gordon Thorburn and Larry Cotton. Thorburn became intrigued with making historic replica furnishings thanks to his friendship with a local furniture craftsman, who was trying to age imitation Jacobean joint stools by soaking them in a soup of, shall we say, organic waste. You don't have to go that far, though—simply building the stool might be enough to prove your dedication to the craft. Cotton wanted to see what kind of chair he could craft from a single piece of plywood. You'd be surprised by how much wood that really is and by the Adirondack-style chair you can make from it.

The art of tying together simple structural members with string, cord, vines, or animal tendons dates back into prehistory. It's called lashing, and it's still in use today. Author Gever Tully's project shows how you can use this ancient art to create sturdy, long-lasting structures out of local materials. Of course, pioneers—not to mention Neanderthals—did not have nylon twine to lash together their structures. But as

we said above, this is a book of old projects done in a modern way.

We moderns have the luxury of preparing and cooking most everyday dinners within an hour or so of when we want them. Pioneers would have had to start the evening meal early in the day, perhaps right after finishing the morning repast. Often they used preserves: vegetables and fruits treated against rot by a process that more or less left them intact and tasty for eating later on. One of these processes is pickling: immersing and storing fresh produce in a solution of spices mixed with salt water or vinegar called a brine, which kills bacteria and leaves behind a tart, crunchy vegetable that does not easily go bad. For this book, author Kelly McVicker shows you how to bring the art of pickling to your own fresh veggies.

The problem of rotting food is even more pronounced with meat. To ensure that they would have a Thanksgiving turkey during their journey westward, some American pioneers carried a brined turkey in a barrel. Brining preserved the meat, made it easier to cook—and by some accounts made the resulting roast bird more delicious, too. So even though you are not riding a wagon that's being dragged by oxen across thousands of supermarket-less miles, you'll want to take on author Katie Goodman's project for brining your own Thanksgiving turkey. (Goodman also supplied us with the chapter on roasting pumpkin seeds.)

When the evening meal is over, it's time to relax.

Many pioneers devoted their free time to making tools that would make their lives easier. Alan Federman's project, building a Da Vinci mechanism that turns rotational energy into an up-and-down motion, is exactly the kind of thing a pioneering tinkerer would work on when the day was done.

This book ends as it began: with a prehistoric project centered on fire. In this project, William Gurstelle (who has had an article in every issue of *Make:* magazine almost since the beginning) shows us how to make a simple oil lamp, to keep the night at bay.

Conventions Used in This Book

 This element signifies a general note, tip, or suggestion.

 This element indicates a warning or caution.

Safari® Books Online

Safari Books Online is an on-demand digital library that delivers expert content in both book and video form from the world's leading authors in technology and business.

Technology professionals, software developers, web designers, and business and creative professionals use Safari Books Online as their primary resource for research, problem solving, learning, and certification training.

Safari Books Online offers a range of plans and pricing for enterprise, government, education, and individuals.

Members have access to thousands of books, training videos, and prepublication manuscripts in one fully searchable database from publishers like Maker Media, O'Reilly Media, Prentice Hall Professional, Addison-Wesley Professional, Microsoft Press, Sams, Que, Peachpit Press, Focal Press, Cisco Press, John Wiley & Sons, Syngress, Morgan Kaufmann, IBM Redbooks, Packt, Adobe Press, FT Press, Apress, Manning, New Riders, McGraw-Hill, Jones & Bartlett, Course Technology, and hundreds more.

For more information about Safari Books Online, please visit us online.

How to Contact Us

Please address comments and questions concerning this book to the publisher:

> Make:
> 1160 Battery Street East, Suite 125
> San Francisco, CA 94111
> 877-306-6253 (in the United States or Canada)
> 707-639-1355 (international or local)

Make: unites, inspires, informs, and entertains a growing community of resourceful people who undertake amazing projects in their backyards, basements, and garages. Make: celebrates your right to tweak, hack, and bend any technology to your will. The Make: audience continues to be a growing culture and community that believes in bettering ourselves, our environment, our educational system—our entire world. This is much more than an audience; it's a worldwide movement that Make: is leading—we call it the Maker Movement.

For more information about Make:, visit us online:

> Make: magazine: *http://makezine.com/ magazine*
> Maker Faire: *http://makerfaire.com*
> Makezine.com: *http://makezine.com*
> Maker Shed *http://makershed.com*

We have a web page for this book, where we list errata, examples, and any additional information. You can access this page at *http:// oreil.ly/1FfLJXV*.

To comment or ask technical questions about this book, send email to *bookquestions@oreilly.com*.

PART I
Morning

Think You've Mastered Fire? Make and Use a Bow Drill.

1

By Wendy Tremayne

While visiting New York's Berkshire Mountains region last winter I happened upon a crew of girls grounded in Earth knowledge. They study with naturalist Michelle Apland at the Flying Deer Nature Center. The group's treasure chest of skills includes how to make bone tools, identify plants and insects, build shelter, create a natural spring, distill water, and whip up a bow drill to make fire without a match.

On a below-freezing day in March, the girls and I traversed a frozen pond. Midway across, the sound of cracking froze me in place. A 12-year-old turned her head over her shoulder to assure me that cracking is no sign of weakness in the pond's ice sheet.

Later, all the girls unexpectedly dropped to their bellies, extended their limbs to disperse their weight on the ice, and reached out for cattails growing beyond the watery edge—the booty of the exploration.

After the harvest, the formation was uniformly reversed, and we set out for a fort that the girls had built earlier that day.

Here awaited a tinder bundle, a bow drill, and an ice half-wall. The wall surrounded a fire pit that was easily lit, without matches, immediately providing warmth.

I hoofed my way out of the woods feeling entirely pooped while the girls carried out an impressive list of things they were going to do before sundown. I took with me this gem of a DIY.

Materials and Tools

- Branches
- String
- Knife
- Handsaw (optional)

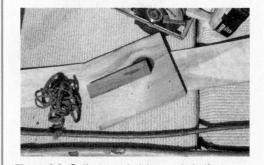

Figure 1-1 *Gather wood, string, and a knife.*

Make a Bow Drill

A bow drill has 4 parts: the bow, the spindle, the fireboard, and the handhold.

Step 1: Choose and Whittle Your Wood

For your fireboard, find or whittle a flat board as thick as your thumb and bigger than 3″ wide and 3″ long.

For your spindle, choose a straight-grained branch that's a thumb knuckle in diameter and 8″ in length. Whittle into a dowel shape, removing the bark, with a semi-sharp point at each end.

Figure 1-2 *Whittle a spindle with semi-pointy ends.*

For your handhold, find a 5″-long piece of wood (slightly narrower in width), semi-flat on one side. To make the handhold notch, whittle a 1/2″-deep hole sloping at 45° to its center.

Figure 1-3 *Notch the handhold to seat the spindle.*

Step 2: Make the Bow

Pick a bow from a live branch that's the thickness of your index finger and a few inches longer than your arm. Choose a string thicker than a shoelace and 5″ longer than your bow. Using a clove hitch or square knot, tie it at each end of the branch.

Figure 1-4 *Tie a thick string to the bow ends.*

Use a Bow Drill to Make Fire

Step 1: Drill a Pit into the Fireboard

Loop the bowstring once around the spindle. With one end of the spindle on the fireboard and the other in the notch of the handhold, apply firm downward pressure to the handhold while bowing back and forth so the bow turns the spindle in both directions (Figure 1-5). Increase the speed. Stop when you've burned a pit into the fireboard that's the diameter of the spindle, and dark with charred wood dust, as seen in Figure 1-6.

Figure 1-5 *Drilling stance (for right-handed people): left foot on the fireboard, right knee on the ground, bow with your right arm, and brace left arm around left knee. (Lefties, reverse the stance)*

Step 2: Make a Notch in the Fireboard

On your fireboard, cut a pie-shaped V-notch that reaches the center of the pit you just drilled. This is easiest if you've drilled your pit near the edge of the fireboard. (You can see the V-notch in Figure 1-7.)

Figure 1-6 *Drill a scorched pit into the fireboard.*

Step 3: Spin a Hot Coal

Place the fireboard on a surface that will catch the black sawdust, which creates a hot ember or coal. Look for smoke and a red glow as you use the bow to spin the spindle (Figure 1-6). When the dust reaches 800°F, it will create a glowing coal (Figure 1-7) that can be placed in a tinder bundle or kindling teepee to create a matchless fire.

Figure 1-7 *V-notch it, and drill again to make a hot coal. Fire!*

Wendy Tremayne is interested in creating a de-commodified life. She was a creative director in a marketing firm in New York City before moving to Truth or Consequences, New Mexico, where she built an off-the-grid homestead with her partner, Mikey Sklar. She is founder of the non-profit, textile-repurposing event Swap-O-Rama-Rama, which is celebrated in over 100 cities around the world; a conceptual artist; event producer; yogi; gardener; backpacker; and writer. She has written for CRAFT's webzine, Make, and Sufi magazine and, with Mikey Sklar, keeps the blog Holy Scrap. She is author of The Good Life Lab: Radical Experiments in Hands-On Living (Storey Publishing, June 2013). Publisher's Weekly named it best summer read for 2013, and the book was awarded the 2014 Nautilus Book Silver Award for Green Living/ Sustainability.

Hogwash: Bacon Soap

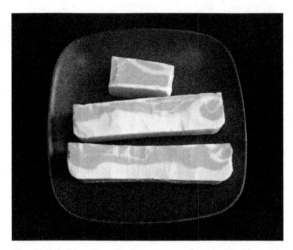

2

Figure 2-1 *How to make bacon soap, from actual bacon!*

By Tim King, PhD

I wanted to see if it was possible to make soap from bacon fat. The bonus challenge: to make the soap look like bacon.

Materials and Tools

- Bacon fat
- 100% lye (sodium hydroxide)
- Purified water and ice
- Liquid smoke flavoring (optional) for fragrance
- Red food coloring (optional)
- Stove
- Pyrex™ baking pan
- Chemical/solvent/heat-resistant plastic bowl
- Measuring cup
- Cotton cloth or paper towels
- Wooden or stainless steel spoon
- Metal can
- Kitchen thermometer

Use the following volume ratio:

- 7-1/2 parts bacon fat:1 part crystalline lye:2parts water

Step 1: Melt Over Low Heat

Go slowly and don't allow it to boil or sizzle. Skim off any particles or debris that float to the surface of the melted bacon fat.

Figure 2-2 *Melt the bacon fat (yum!) over a stove.*

Step 2: Filter

Filter the melted bacon fat by pouring it through cloth or paper towels into a large, clean metal can.

Figure 2-3 *Filter the melted fat, to ensure Ivory Soap™ purity.*

Step 3: Measure

Still warm, your bacon fat should look uniform, clear, and junk-free. Measure your bacon fat, then calculate the proper amount of lye—1 part lye to 7-1/2 parts bacon fat.

Figure 2-4 *Measure lye crystals (wear eye protection).*

Step 4: Pour

Measure 2 parts purified water. Feel free to include some ice in it, as the lye will get very hot. In a heat-resistant, nonmetal container, pour in your 2 parts water, then slowly pour the crystalline lye into the water.

Warning!

Lye fumes are noxious and can cause serious injury. Make sure to wear rubber gloves and eye protection when handling lye, and to mix it in a well-ventilated area. Also, the lye and water mixture will get very hot, so be sure to mix them in a proper workspace, and in a heat-resistant, nonmetal container.

Step 5: Wait

Take a break and wait for the lye-water mixture to cool down. Do not leave the lye mixture unattended. When the lye-water mixture is around 100°F (the temperature of a hot shower), check your bacon fat and see if it's around the same temperature (if not, heat it on the stove until it is). Then slowly pour the warm lye mixture into the warm bacon fat.

Figure 2-5 *Pour warm lye mixture into the warm bacon fat.*

Step 6: Stir

Stir the mixture constantly for up to two hours as it cools. When the mixture begins to show "trace" (white soap stripes on the top surface), move on to the next step.

Figure 2-6 *Stir the mixture until you see trace (white streaks) on the surface.*

Note

Mixing can take several hours. If your mixture is still liquid after 2 hours, move on to the next step and hope for the best.

Step 7: Pour (Again)

While the proto-soap mixture is still a thick liquid, but showing signs of trace, you can mix in liquid smoke flavoring as a fragrance to enhance the bacon smell. Also, if you want to make your bacon soap look like bacon, now's the time to pour a small portion of the mixture into a separate container and add a bit of red dye. Pour the beige, thick proto-soap into a pyrex glass pan, then take the red-dyed proto-soap and pour stripes into it to make it look like bacon.

Figure 2-7 *Add red dye (optional) to a small portion, to honor the original bacon.*

Step 8: Cure

Let the soap set or "cure" for at least 36 hours, then cut blocks into desired shapes from the greater slab. Do this while it's still relatively soft. Allow the soap to continue to cure for at least two weeks.

Figure 2-8 *Cut the partially cured soap into desired shapes.*

Step 9: Enjoy!

Enjoy your soap. Everybody loves bacon (the meat of the gods), and now we can bathe with it.

Dr. Timothy King is an archaeologist and an anthropologist. He enjoys reproducing ancient technologies and he excavates Ice Age mammoths in his free time.

Kitchen Table Cider Making

3

By Dr. Nevin J Stewart

Yes, our new neighbor got as drunk as a skunk. Blootered, bladdered, and blitzed as we Scots say or, in cockney slang, Brahms and Liszt! He and his partner came along to a cider-themed evening at our house where he was initiated into the Scillonian Road hard cider making cooperative.

We started the evening with a hands-on session of how to use my "juice and strain" method to make clear apple juice and crystal-clear hard cider, quickly and with minimal mess. Rather than smash the apples outdoors with a messy, old-fashioned cider press, I use modern centrifugal juicers right on the kitchen table.

Dinner was accompanied by last year's homemade hard cider (6.5% alcohol by volume), apple wine (15% ABV), and store-bought Calvados (40% ABV). It was the latter that did him the most damage.

The next day, our new friend could remember nothing of the evening's proceedings. He was unable to recall disclosing his lifetime's accumulated prejudices concerning ethnic minorities, politics, religion, and relationships. Also my introduction to the juice and strain cider method had been lost, as with the spirit vapors.

To save further embarrassment, I said nothing about his verbal indiscretions, but I did re-explain my method to him as follows.

Materials and Tools

- 75 lbs apples, twice washed
- 5 g packet Champagne-variety yeast (Saccharomyces bayanus)
- Campden tables (for sanitizing; they can be found at wine-making and home-brewing stores)
- Sugar (optional, for sparkling cider)
- Whole-fruit juicer, such as the Breville Juice Fountain Elite or Plus; buy the highest-power machine you can afford

- Food-safe plastic hose to fit juicer spout; typically 1" ID × 16" long
- Cotton towels
- Spring clamp
- Fine straining bag, typically 24" × 24"
- 5 gallon brewing bucket, with open top, with tap at bottom typically 12" diameter × 17-1/2" tall
- Two 2-gallon pails For sterilizing and straining. The straining pail should be the diameter of your brew bucket, or a bit larger; drill 30+ holes in the bottom.
- Hydrometer
- Funnel
- 5-gallon carboy, or five 1-gallon demijohns
- Airlocks(s) and rubber stopper(s), which adds about 5" to your demijohn/bucket height
- 2-cup measuring cup

For Bottling:

- Rubber tubing for siphoning
- Siphon tap (optional)
- Tray with rim, such as a cookie sheet
- Beer-type bottles
- Bottle caps, crown style
- Hand-operated bottle capper
- Labels

Step 1: Clean Your Equipment

Apple juice and hard cider are foodstuffs, and all appropriate food handling and safety measures should be stuck to. Wash your hands, sanitize all surfaces, double-wash the apples, and throw away any bad ones.

Sterilize all equipment that will be in contact with fresh apple juice. I use a stock solution of four Campden tablets per gallon of water to soak all the relevant parts and buckets for a couple of hours before use.

 Warning

If you're making only apple juice (and not fermenting it into cider), be sure to use only handpicked apples, as windfalls may have picked up harmful bacteria which might not wash off.

Step 2: Set Up the Juicer and Strainer.

Lay out a clean towel, rinse off the juicer parts, and assemble your whole fruit juicer.

Attach the "juice containment and delivery adaptor," (a.k.a. hose), to the juicer's spout, and feed it into the straining bag, held within a straining bucket that has holes in its base. This assembly sits neatly in the open brewing bucket with a draw-off tap at the bottom.

Set up your brew bucket on a stool or box, high enough that you can fit your demijohn or carboy underneath the tap. Apples go in at one end, clear apple juice comes out at the other. It couldn't be simpler.

Step 3: Juice and Strain.

Feed apples into the juicer with a steady, even pressure on the pusher. The higher the machine's power rating, the faster you can go.

Warning

Whole-fruit juicers are powerful appliances. Read and adhere to the safety and other instructions for your juicer.

When the pulp container fills up, discard the pulp. After every 25 lbs of fruit, dismantle the machine and clean the pulp off the centrifuge stainless-steel mesh.

At this point, measure the original gravity (OG) with a hydrometer and write it down. Later, this figure will allow you to estimate your cider's alcohol percentage. If the OG is low, top it up with a little white sugar to reach 1.040.

You'll find that the juicing work is done in a flash, although it takes a while longer for all the juice to strain through. I obtain the last 5% of the expected 65% by weight of juice by wringing out the straining bag. Scottish, you see!

What's left in the bag is about 1% of the original apples. This very fine pulp can be used in apple muffins. You don't want it in your cider.

Step 4: Shoot the Yeast

While the last juice is draining, pitch the yeast into a measuring cup containing fresh, clear apple juice held at room temperature. This will allow the dried yeast to rehydrate and kick-start your fermentation. Use a champagne yeast for simplicity and reliability. A 5 g packet is enough to inoculate 5 gallons of juice.

After half an hour, stir the cup to thoroughly disperse the yeast, then pour it into your sterilized carboy or demijohns. Fill these up nearly to the top with apple juice, and put airlocks on top. Within the hour, you should see bubbles coming out through the airlock.

When the hard cider is finished, measure the final gravity, and read off the alcohol content from an ABV chart or online calculator. For reasonably good storage, 5% ABV is considered the target minimum.

Step 5: Ferment

Keep the fermentation vessel(s) in a warm place like the kitchen: and after three weeks, you should have a crystal-clear cider ready to be racked and bottled. Check it with your hydrometer. The reading needs to be 1.000 or less. If it's still high, let fermentation continue.

Step 6: Bottle

Siphon your cider into recycled, sterilized beer bottles that will take a crown cap.

If you want a still hard cider, just bottle as is. If you want bubbles, then add 1/2 teaspoon of white sugar to a pint bottle, fill up with your hard cider, and cap. After a few more weeks, a secondary fermentation should be complete, and you'll have some fizz.

Step 7: Enjoy

Serve chilled. Take care when opening. If you've overdone the sugar, it can go off like a fire extinguisher.

When serving, you can adjust the sweetness to taste by adding sugar syrup. But I prefer my cider as dry as it can go.

Nevin Stewart, a retired industrial chemist living in Guildford, England, no longer "cooks up" novel oilfield production chemicals for British Petroleum. Instead he has trained as a hobbyist chef and creates new fusion dishes for family and friends. These are usuallly paired, and washed down, with homemade apple juice or hard cider.

Paper Making

—Recycle your unwanted bills and junk mail into custom-made personal stationery.

By François Vigneault

Paper, first invented in China around the first or second century AD, is now so ubiquitous that it has achieved near-invisibility in our modern world. The average American household receives more than 100 pounds of unwanted junk mail each year! However, creating a handmade sheet of paper can remind anyone of this everyday object's noble origins.

Here's a chance to give your unwanted papers a second life for stationery, collage, or anything else you can imagine. Rediscover this ancient and oh-so-easy art form.

Materials and Tools

- Sponge
- Interfacing or wool blankets for felts
- Old paper products: old bills, junk mail, and other scrap papers, and inclusions (optional), such as fabric or paper scraps, glitter, leaves, or flower petals
- Cookie cutters (optional)
- Cookie sheet or other surface for water control
- Brayer or smooth flat block.
- 1" x 4" lumber cut into two 5" pieces and two 9" pieces.
- 4 Hook-and-eye latches
- Canvas stretchers, sold at art supply stores
- Window screening
- Rubbermaid™ tub or other waterproof container
- Blender
- Hammer
- Staple gun

Step 1: Learn the Glossary of Terms

Mold and deckle:

The frame that's used to make paper. The mold is the bottom portion, which includes the stiff mesh that the screen rests on. The deckle is the upper portion, which determines the shape and size of the sheet of paper (the ragged edges seen in handmade papers are called deckle edges). There are many different versions of this mold-and-deckle setup; for this project, we use a variation called a deckle box or pour mold.

Couching:

Pronounced "cooching" (it's derived from the French coucher, "to lay"), this is the process of transferring the wet sheet from the mold to another surface (the felt) to dry.

Felts:

Sometimes called couching sheets, these are the pieces of material used to separate sheets of wet paper while they dry. Felts are available at art supply shops; however, interfacing or old wool army blankets can also work well.

Pulp:

The mix of water and plant fibers that your paper is made of. You can create pulp from cotton linters (sold at art supply stores) and other plant fibers; but in this project, we'll make pulp by reusing scrap paper and junk mail.

Step 2: Build the Pour Mold

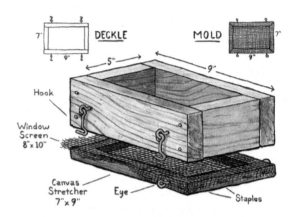

Cut the window screening to a size slightly larger than the outside dimensions of the canvas stretcher (my canvas stretcher here is 7" × 9"). Use a staple gun to attach the window screening mesh to the canvas stretcher, making sure it's as taut as possible.

Cut 4 pieces of 1" × 4" lumber (here my pieces were 5" and 9" long) to fit together into a rectangle—the interior size will be the finished size of your paper (5" × 7" in this case). Secure the sides with wood glue and nails.

Add the hook-and-eye latches on either side of the mold and deckle box to hold them together tightly.

Step 3: Make the Pulp

The paper you choose to recycle will affect the consistency, color, and feel of your handmade paper. In general, bills and printer papers will create a smoother, more consistent sheet, while magazine pages and glossy papers will tend to chop up more irregularly, creating a more "artistic" look. Experiment with mixing different papers together.

Cut or tear your paper into approximately 1" square pieces, and place them in the blender with enough water to cover them completely. With the pour method, you can make as much pulp as you want at a time; a good rule of thumb is that whatever the size of your original sheet, the pulp you make from it will make a sheet about 1" smaller in both width and length.

Blend the paper scraps and water until all large chunks are pulverized (about 30 seconds to 1 minute). The longer you blend the pulp, the smoother and more regular your paper will be. Pulping can dull your blender's blades quickly, so it's a good idea to keep a dedicated paper-making blade or get a separate blender (you can usually find one at a thrift store) if you want to make paper frequently.

Personalize your pulp! You can add in a wide variety of materials while blending, including leaves, flowers, glitter, confetti, seeds, and much more. It's best not to blend ribbons or other long fibers, as they can get wound around the blades. Finally, consider dyeing your pulp.

Step 4: Pour the Pulp

Fill your vat with enough water to cover the mesh on the resting pour mold by at least 1/2". (A large Rubbermaid™ tub works great, and can be used to store your papermaking supplies when not in use.)

Using the hook-and-eye latches, secure the deckle box and the mold. Place them in the vat, sliding in at an angle to discourage air pockets from forming.

Pour your pulp onto the mesh. The more pulp poured, the thicker the paper. Use your fingers or a spoon to stir the pulp and distribute the fibers evenly across the surface of the water.

Slowly pull the pour mold upward, letting the water drain back into the tub. Place the entire apparatus on a cookie sheet to keep water from getting everywhere.

In traditional Western papermaking, the entire vat is filled with a pulp mixture, which makes for more consistency from sheet to sheet. The pour method, on the other hand, makes it really easy to vary the texture, color, and weight of your sheets; each one can be totally different! This makes it an ideal technique for beginners, who may want to experiment wildly before settling on a style.

Step 5: Couch the Sheet

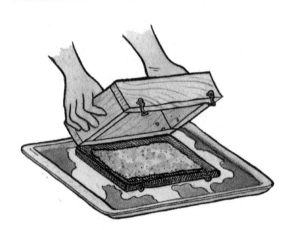

Undo the hook-and-eye latches and lift the deckle box, being careful not to pull the wet sheet up with it.

Lay your felt (I used interfacing for this project) onto the wet sheet. Carefully turn over the felt, sheet, and mold. Be sure to hold the layers together.

Step 6: Use the Sponge

Use a sponge to soak up any excess water from the sheet, pressing down on top of the felt and wringing out the sponge until you can't pull any more water out of the paper.

Place another felt piece on top of the sheet. Using a brayer or presser bar (or any smooth, even surface), smooth the paper to remove any lingering excess moisture.

Step 7: Dry the Paper

Slowly lift the mold from the paper surface, holding down the felt. The surface tension between the felt and the paper is greater than that between the paper and the mesh, which should cause the paper to stick to the felt.

Set the paper between the felts on a flat surface to air-dry. Your paper may "cockle" (curl) a bit; if you want to reduce cockling, stack your wet sheets, one on top of the other, with felts between each sheet, then place a heavy book on top to press them.

The drying time of your papers will vary from less than an hour to several days, depending on the humidity in the air and the type of pulp used. If the drying takes more than a day, change the felts once a day—this will keep the paper from getting moldy. If you're in a hurry, you can gently press your sheets with an iron, but this tends to make the sheets cockle quite a bit, and I don't recommend it.

Step 8: Variations: Paper Shapes

You can easily incorporate designs into your paper by separating different pulp colors or textures into simple designs.

A piece of stiff, thin plastic can split your sheet into two or more sections. You can also use cookie cutters or a tin can with both ends removed (for circles, as seen in the opening shot) to create shapes within your sheet, or to make shaped gift tags, etc.

Pour distinct pulp mixtures into the separated areas, pull your mold from the vat, and remove the separators before couching your sheet. The pressure from the felt will join the separate sections into a single sheet, as long as they're approximately the same density and weight.

Step 9: Variations: Embedded Items

If you'd like to embed flat items such as paper, fabric, or leaves into your paper, it's easy. Dip the item into your pulp mixture to coat it with a thin layer, and then work the item into the pulp sheet right after you remove it from the vat.

François Vigneault is a freelance illustrator, the creator of the sci-fi comic Titan, and a cofounder of the Linework NW festival. He lives in Montréal, Québec.

Olde-School Bookbinding

5

—Pages last longer, lie flatter, and look better inside a handsome, durable hardcover.

By Brian Sawyer

Magazines aren't really built to last, but here's how you can turn your copy of *Make:* (or any other magazine or printout) into a durable hardcover that will withstand the test of time. Your hardcover *Make:* will also lie flat on your workbench, making it easier to follow instructions for other projects.

Tools and Materials:

- Magazine or pages to be bound
- Acid-free nontoxic adhesive (I'm a Yes man, myself)
- Bookbinding tape, 1/4" linen
- Durable, acid-free linen binder's thread
- Decorative paper for cover and end sheets
- 1/8"-thick binder's board, or use chipboard or illustration board
- Loose-weave, white linen fabric
- Heavy-duty ruler or carpenter's square
- Utility knife
- Pen
- Binder's needle or tapestry or other heavy needle
- Brush
- Scrap paper
- Wax paper
- Medium-grit sandpaper
- Medium-duty awl

Step 1: Create the Signatures

Peel away the existing cover, and use a utility knife and a heavy ruler to cut out all the pages 1" from the spine, freeing them from the glue. Divide your loose pages into consecutive 32-page sections (signatures). Binding loose pages as joined signatures will strengthen the spine and keep pages from falling out. For *Make:* Vol. 01, without the ads, I got 192 pages, or six groups. If your total page count isn't a multiple of 32, you can fudge the signature sizes a bit, but each signature must have a page count divisible by 4.

Open the first group in half, such that pages 16 and 17 are facing. Pair these pages and set them aside, doing the same for the next facing pages (14 and 19) and the rest in the group. Just keep subtracting 2 from the left side and adding 2 to the right to determine which sheets to pair up. If you did it correctly, the last two pages you pair up will be 2 and 31.

Draw a vertical line 1/4" from the right edge of page 16 (and all up-facing even pages). Cover the area to the left of the line with scrap paper. Brush a thin coat of acid-free, nontoxic adhesive (such as Yes) into the exposed 1/4" gulley. Press the corresponding up-facing odd pages into the gully to glue together the pairs.

Return the pairs to the order they were in before being split in half. Align the edges of the pages and fold them along the spine. Repeat for each group and collate the finished signatures in their original order.

Step 2: Stitch the Signatures

Sewing your signatures together around bookbinding tape creates the added durability of a hand-bound book. Measure and make two marks along the fold of the first signature, 1/2" in from each edge of the signature. These two marks represent the kettle stitches, the stitches that connect one signature to the next.

Figure 5-1 *Ready to sew. Though using a stitching post is not absolutely necessary (as long as you keep a steady hand, make sure the tapes remain taut, and ensure that the signatures stay even while you sew), it does keep the work organized and easier to manipulate with the only two hands you have. If you have one available, set it up with the tapes stretched tight and spaced to match your holes.*

Now measure and mark six more points on the fold: three pairs of points, 1/4" apart, spaced evenly between the two kettle stitches. These represent the in-and-out point for sewing the signatures around three tapes, which will run behind the signatures.

Stack the remaining signatures and make the same marks, at the same measurements. Pierce the marks with an awl, making holes just wide enough to allow a needle to pass through snugly.

Using a heavy needle, enter the spine and pull about 30" of thread through the foot (bottom edge) kettle stitch of the last signature. Exit the spine at the next hole, and re-enter around the first tape.

Figure 5-2 *Attaching the signatures with three strips of bookbinding tape.*

Keep stitching around the tapes, and exit the spine at the head (top edge) kettle stitch. Using the same thread, enter the next-to-last signature at its head kettle stitch. Stitch around the tapes, and knot the thread around the foot kettle stitch of the first signature.

Continue in this fashion to stitch the remaining signatures together. If you run out of thread, knot a new 30" length to the existing thread. The best place to do this is just before re-entering the spine around a tape. When you come to the last kettle stitch, knot the thread.

Figure 5-3 *Use a heavy needle to penetrate the paper. The three tapes (one is shown here) run behind the signatures.*

At this point, it's a good idea to apply a bit of glue (about 1/4") to the inside of the first and last signatures (use a piece of scrap paper to protect the portion of the page you don't intend to glue) and put the work under heavy weights overnight.

Step 3: Glue the Spine

Next, attach the mull (a strong strip of cloth with a loose weave that allows paste through it) to the spine and tapes. (Connecting the cover boards to the mull, rather than directly to the signatures, allows for a flexible backbone. This is the key to lay-flat binding.) Keeping the pages aligned on all sides, sandwich your work in a press or vise.

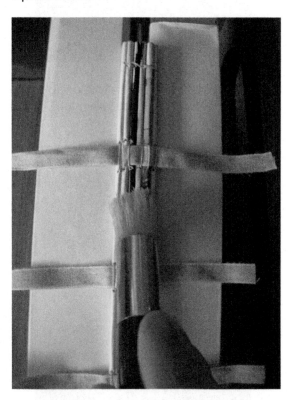

Cut a piece of mull that's tall enough to cover your kettle stitches and 3 inches wider than the width of the spine. Brush a generous amount of glue on the spine, from kettle stitch to kettle stitch, across the full width. Lay the mull flat and mark the spine area on it, 1" in from either side. Brush this area generously with glue.

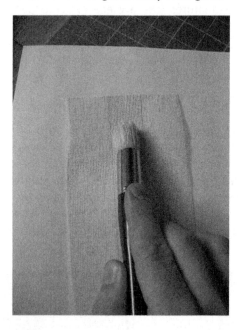

Position the mull symmetrically and rub it into the spine. Leave the book overnight under heavy weights or in a press.

Step 4: Attach Cover Boards

Cut three boards (1" binder's board) for the front cover, back cover, and spine. Allow 1" additional clearance on the head, foot, and fore edges. Reduce the edge on the spine side of the front and back covers by 1/4" (the thickness of two boards), to accommodate the hinge. Altogether, the covers should be cut to 1/4" taller than the height of the book (1" added to the head and foot) and 1" narrower than the width of the book (adding 1" and subtracting 1/4").

Cut the spine to the same height as the covers and the same width as the signatures. Sand down the rough edges.

Place one piece of wax paper between the mull and the free ends of the tapes and another beneath the tapes. Brush the free mull edge with glue. Remove the top piece of wax paper. Press the front board against the mull, extending 1" of the board over the head, foot, and fore edges.

Open the cover and rest it against a board for support. Rub the mull with cloth or paper, working the glue into the board. Brush the tapes with glue and press them to the cover board. Discard the second piece of wax paper, and place another clean piece between the cover board and the first signature. Repeat for the back cover.

Step 5: Cover the Cover

You'll now cover this skeleton with a single piece of decorative paper that wraps around the front cover, back cover, and spine. To create room to slide the paper over the edges of the cover and spine, use your utility knife to slit the mull by 1/2" where the covers meet the spine, at both the head and foot edges.

Lay your cover paper face down and mark the placement of your boards. Allow a 1/2" (thickness of four boards) turnover width for all edges, and 1/4" (thickness of two boards) for each hinge.

Brush the spine area of the paper with glue, position the board, and press firmly. Turn the paper over and rub to secure the spine, and mold the paper over the edges of the board.

Brush the area you've marked for the front cover with glue, brush slightly into the turnovers and hinge, and press the board between your marks. Turn the book over, and rub the cover to remove air bubbles or wrinkles.

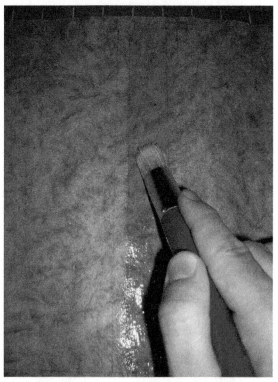

Figure 5-4 *Brush the area you've marked for the front cover with glue.*

Brush glue onto the area for the back cover. Lay the back cover board's fore edge down on the paper, meeting your mark for that edge. Pinch the paper to the board on that edge, and press the remaining paper to the back of the book.

Rub the back cover, working the paper on into the hinge to seal the paper to your book block around the spine. Repeat for the hinge of the front cover.

Figure 5-5 *Working the paper into the hinge.*

Lay the book open, and brush glue across the length of the head turnover. Stand the book up on its foot edge, and roll the edges of the head turnover over the top of the board. Repeat for the foot edge. Brush fore edges with glue, fold the turnovers, and smooth out wrinkles.

Figure 5-6 *Smooth out wrinkles.*

Lay sheets of wax paper between the cover boards and your block of signatures, and press under heavy weights overnight.

Step 6: Finish

Pasting end sheets to the inside front and back covers reinforces the spine and finishes your book's appearance. I used the original *Make:* wraparound cover as my front end sheet. Trim it to leave an equal distance around each edge, and paste it to the board. Allow glue to run into the spine, covering the point where the cover meets the spine and extending into the first signature by a 1/4" gutter. Repeat for the back cover board.

Put fresh pieces of wax paper between the covers and the book block, and then set under heavy weights to dry overnight. You'll wake up to a long-lasting volume that will look unique, lie flat, and serve you well.

Brian Sawyer makes books, makes stuff, and makes books about making stuff. A bibliophile at heart, and a book editor by trade, he often writes about making books on his personal blog, when he's not busy actually making them.

PART II

Afternoon

Fool's Stool

Figure 6-1 *Build a fake colonial period stool good enough to fool almost everybody.*

By Gordon Thorburn

Almost anybody can make A 17th-century board stool. In those early, pioneering times, techniques and tools were fairly primitive and ambitions consequently modest, so today, faking the American colonial style requires a do-it-for-fun, cavalier attitude rather than serious precision.

In making your brand-new antique, it actually helps if you begin with joinery that's not perfectly straight, level, smooth, and right-angled.

You're not making a direct copy. The photographs are here for guidance, but each fake piece is an individual article. It's not so much a clone as a close relation of something genuine.

When British settlers turned up in America, they brought the Jacobean style with them and translated it as American colonial, but they couldn't go to the wood merchant and buy planed and sanded oak, cut for them in convenient widths and lengths. They didn't have an electric jigsaw or a power drill, but you probably do. You'll also need a hammer and a few odds and ends. Compared to the Pilgrim fathers, you have it very easy.

Materials and Tools

- Oak boards, 18" – 20" long:
  — one board, 12" – 14" × 3/4"
  — two boards, 10" – 12" × 1/2"
  — two boards, 8" × 1/2"
- Pine scrap, about 1" × 1" × 4"
- Paste finishing wax and wood glue
- Cow manure (preferred) or wood stain and shellac
- Drill
- Jigsaw
- Handsaw
- Circular saw

- Hammer
- Chisel
- Spokeshave
- Brush (optional)

Step 1: Cut the Five Pieces of Wood

We're not going to give precise dimensions, because they don't matter. You have five pieces to cut: a top, two identical legs, and two identical sides. Draw templates on paper, and pencil in the outlines on the wood.

Figure 6-2 *Pencil the template outline onto the wood*

The height of the stool is about the same as the length of the top. The top is between 18" and 20" long, with grain running longways, 12" to 14" wide, and 3/4" thick.

Legs are about 2" narrower than the top, to give the overhang. Sides are about 8" wide. Decorative cuttings-out with your jigsaw must be kept simple. You could just drill a few large holes in a pattern.

Figure 6-3 *The 4 pieces of the frame are joined by 4"-long, 1/2"-wide slots*

To obtain the squared-off end on the slots, drill a hole in each corner of the slots so the jigsaw blade can make the 90° angle.

Figure 6-4 *Drill a hole to help the blade*

The slots in the sides should be at a slight angle. If the side is 18" long, the gap inside the legs at the top edge should be about 12" and at the bottom edge about 13-1/2".

Step 2: Distress the Pieces

The fate of most of your stool's rustic and very early brethren was to be chopped up and put on the fire when the more elegant stuff came in. For it still to be here in the 21st century, it must have suffered in many ways, to be rescued at last by your good self or another saintly person you know.

Before assembling the stool, rub with coarse sandpaper and throw and drag your cut pieces of oak around the yard or along the street until you get a satisfactory number of chips and scratches.

Figure 6-5 *Distressing the oak*

Step 3: Glue the Pieces Together

The colonial maker would have used nothing but joinery to hold the frame together, but then his stool was only going to be sat on. His stool wasn't going to be thrown around, soaked, shot at, and otherwise attacked (see steps 5 and 7). You shall use wood glue.

Figure 6-6 *Use wood glue to hold the frame together*

Depending on the accuracy of your slot cutting, you may need to tie string around the leg pieces to keep them in order while the glue sets.

Step 4: Fix the Top

Drill 1/4″ holes through the top into the legs, one at each corner, and two more at the mid-point into the sides. Do this by eye. Don't measure.

Figure 6-8 *Square pegs to be fitted into the round hole you just drilled*

Figure 6-7 *Doing it by eye makes it look more authentic*

Use a chisel and hammer to make eight square pegs out of pine wood. Use a piece of pine that has a nice straight grain to make it easier when splitting. Make square-section lengths, like very fat matchsticks, slightly thicker than the holes you've drilled. If your peg-wood has good straight grain, you can split it with a chisel; otherwise saw it.

Drop some glue in the hole and bang the peg in fearlessly with a hammer. Saw off any surplus, not too close.

Figure 6-9 *Hammer in each peg fearlessly, then saw off the remainder.*

Step 5: Distress the Stool a Bit More

The expert looks for a texture to the wood on the seat, with the grain standing out a little, the pegs proud, and a general air of being used and worn and knocked about. Obvious places for wear are the bottoms of the legs and the edges of the seat. Also, the underside of the overhang should show the patina of being picked up a million times by greasy fingers.

So, how do we fool our friends and neighbors? First, add wear, again with coarse sandpaper and a spokeshave. Simulate dog bites with vise grips. Don't overdo it, but allow no line to be perfectly straight.

Figure 6-10 *Bad dog!*

Step 6: Stain with Cow Muck

Next, it's the color. With oak, there's only one way to get that perfect, 500-year-old look, and that is to make your stool and wait 500 years.

The best compromise between time and convenience is to steep your oak in cow muck. We realize this is not an option everyone can pursue, but then the best never is. The process gives you a very good color that's not flat, as stain tends to be, but variegated in a natural, haphazard way.

You could go and collect a sack full of cow pats, make them into a soupy slop in a trash can, and immerse your stool in it. Add your used tea leaves and coffee grounds. Compost is good. Four to six weeks will be a reasonable time in the pit, but a mid-term inspection will be necessary. The color revealed when wet is the color you'll get when it's polished.

Once you're satisfied with the color, you'll have to get rid of the smell. A week in running water will do it, so it's handy to have your own trout stream or access thereto, or a week in the sea. Otherwise it's a fortnight or more in a rain barrel with regular water changes, lots of wet weather, or the equivalent in garden hose soakings. When the wood is wet, that's the time to add wear to the bottoms of the legs. Tap with a hammer to splay the feet and round them off, to simulate years of scraping on a stone floor.

Drying out in sun and wind between soak-seems to help lose the smell. Don't be dismayed by the light gray it goes when dry. Polish will bring it all back. On no account use any chemicals or proprietary deodorizers. These may spoil your color and/or seal in the smell so it will never go away.

Another great thing about soaking is that you're likely to get the odd random crack or two, which adds an authenticity you could never get on purpose. It also raises the grain as if a hundred family backsides have gradually worn away the softer parts of the wood, leaving the harder grain standing out in tiny, well-polished ridges.

Figure 6-11 *An odd random crack ...*

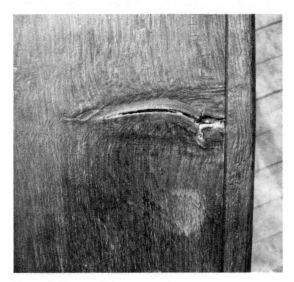

Figure 6-12 *... or two.*

If cow muck isn't for you, you have alternatives, such as stain. This is the last resort because it tends to give a flat result, and it's the hardest to get right.

If you have no other option, then at least do a week's soaking in water, bash the feet, and allow the stool to dry out thoroughly, preferably in the sun.

Mix dark oak stain with shellac, about 50:50 (try more or less to your liking). They don't blend, which is the idea. When you like the result, slap it all over the stool with a brush. Do this at speed, with confidence.

Don't worry about runs and uneven areas. When it's dry, go into the shade and rub a little wax furniture polish on it. Phew. Let's hope it works.

Step 7: Distress the Stool Even More

We give here a menu of possibilities for mistreatment, but be circumspect. No one piece could have suffered all of these miniature disasters. Beware of overfaking.

Use it as a sawbench. Poke holes in it with an awl. Scratch it with a dinner fork. Carve initials on the top in a corner, or at the top of a leg. Draw a rudimentary coat-of-arms with a soldering iron. Knock off a vulnerable bit and glue it back on. Scorch part of it with a blow-torch. Stain it with sloe gin or ink. Burn it with a cigar or cigarette. Some of these techniques will reveal new wood beneath. Black shoe polish will sort that.

Now you can polish. Rub all over with diluted shellac to seal the color, then wax polish except for the main underside, which never saw the light and was never touched. And that's it. Best of luck.

Go on, you can do it.

Gordon Thorburn is the author of over 25 books.

Rok-Bak Chair

7

—Can a chair be comfortable, look good, recline, disassemble for compactness and portability, and still be made from just a single sheet of plywood?

By Larry Cotton

With a determination borne of frustration and frugality, I set about designing a chair that would meet all my requirements. I began by making a crude study model, with all body-supporting surfaces adjustable: the seat, the back, armrests, headrest, footstool, the angle between the seat and back, even the overall size. The model's only given: it would use a set of standard patio chair cushions and one sheet of plywood.

Somewhere in the middle of seemingly infinite adjustments, I discovered that recliners really are much more comfortable than upright chairs, so I threw reclinability into the mix as well.

Finally I found a combination that fits—ergonomically, esthetically, and economically. It's even been sleep-tested. I call it the Rok-Bak chair.

The Rok-Bak is very comfortable, easy and inexpensive to build, can be assembled or disassembled in a few minutes, and can be stored and moved about easily, taking up very little space.

The Rok-Bak can be built in two configurations. One is just a nice, comfortable chair. The other has bottom edges shaped like shallow Vs, and can be rocked back (hence the name) into a reclining position.

Either configuration can be complemented by a headrest and footstool, but in Rok-Bak mode, you'll definitely want to make both, for even more comfort.

If you're really ambitious, you can upholster your own cushions (as I did with expert help from my brother Phil), as well as the headrest and footstool, for a perfectly matched set. The upholstery techniques are similar for most any fabric.

You can cut the chair parts with a jigsaw and a circular saw. Fabric work can be limited to cutting with good scissors and making 1 long seam—with either a sewing machine or fabric bonding tape—then stapling in place.

You'll also need additional hardware and staples. You should decide at the outset which configuration you like: basic or Rok-Bak. Keep in mind that the Rok-Bak is quite usable rocked forward. Once you cut the large bottom cutout, you can't change the chair back to the basic configuration. However, you can later convert basic to Rok-Bak.

Since the *A* side of the plywood (the good side) faces outward on the chair and stool, the *C* side, with its knots and other imperfections, will be almost completely hidden.

Materials for the Chair Only:

- 2 Patio chair cushions, 4"–5" thick stock or custom. Ideally get a 22" × 40" stitched-together pair; alternately, get two separate cushions, each 22" × 20". Big-box stores carry patio cushions, and fancier cushions can be found online in lots of colors and designs for a bit more money.

- Varnish, or, preferably, polyurethane spray

- Spray adhesive for holding foam on arms during upholstery

- 8×1-1/2" flathead wood or drywall screws (22)

- 10 9/32"–5/16" ID washers

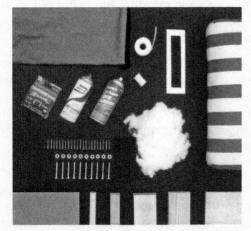

Figure 7-1 *Materials you'll need*

- 10 1/4" × 2-1/2" lag bolts (10)

- 2 × 4 fir or pine, 8' length for cross braces

- 1 × 4 fir or pine, 6' length for arms

- 1/2" plywood, 4' × 8' sheet, A-C interior grade (minimum)

For the Footstool and Headrest:

- Hook-and-loop fastening material (Velcro™), 30 square inches

- Fabric bonding tape such as Stitch Witchery, 1/2"–5/8" wide such as Stitch Witchery. This is an alternative to sewing.

- 1 yard of 54" or 60" upholstery fabric. Material for upholstery is abundant at mill outlets. I used denim for my deluxe chair, which looks good, is easy to work with, and (you guessed it) cheap.

- 1 box staples, with 3/8" leg

- Spool of white thread (optional)

- 4 oz (minimum) pillow stuffing, such as Holofil or Nature-fil, for the headrest

- Medium-density foam, 29" × 19" × 1-1/2" thick, from a foam store or upholstery shop. Medium density weighs about about 1lb/ft^3.

- 1" wood dowel, 8-1/2" length for the headrest

- 2×2 fir or pine, 3' length for footstool top supports

- 2/4" foam, 7" × 25" or you can saw 1-1/2"-thick foam

- Jigsaw with plywood blade

- Circular saw (optional) with plywood blade

- Sander with various grits of sandpaper Random orbit is best.

- Drill with drill bits and screwdriver bits

- 1-1/2"–2" diameter drum-sander accessory for drill with 60-grit paper

- 7/16" ratchet wrench—A manual ratchet is OK, but an accessory for your drill is faster.

- Sharp pencils with erasers

- Standard 15oz can. You may eat the contents, but save the can.

For Upholstering:

- Scissors or shears
- Electric knife (optional) for cutting foam
- Band saw (optional) if you're re-sawing 1-1/2" foam; otherwise buy 3/4" foam
- Staple gun
- Sewing machine or clothes iron for sewing or hot-taping fabric seams

Step 1: Make the Plywood Pieces

Download all the construction drawings at http://bit.ly/rokbakgithub.

Start by using a pencil to lay out the plywood pieces on the *C* side of the plywood. Why? Because we'll cut the parts out with a jigsaw (and possibly a circular saw), and we want to keep the *A* side (think "appearance" side) splinter-free. Both saws' blades cut on the upstroke, so any splintering will be confined to the (mostly hidden) *C* side.

You should lay out only one side of the chair and footstool. Later you'll use it to trace the other side, making sure they are mirror images, as shown in the layout drawing. Use a standard 15 oz can as a radius (about 1-1/2") template for corners. A sharp jigsaw blade should have no trouble following that radius.

Figure 7-2 *Use a can as a template*

To conserve plywood, you can skip laying out the footstool top. It can be pieced together later from the chair-side cutouts.

 Tip

You must cut the top corner radius of the chair side with a jigsaw; a circular saw would nick up the footstool side.

Suspend your plywood on a few supports. Scraps of 4 × 4 wood or even paint cans (if they're the same height) make good supports. Position them, obviously, out of the projected path of the saw blade.

Cut one chair side and one stool side. Start by jigsawing out the cutout under the arm. Drill a 3/8" starting hole from the *C* side—inside of, and close to (but not on), the layout line. Use a scrap of wood on the exit side of the hole to prevent splintering. Then insert the jigsaw blade into this hole and cut as usual.

Note

Both chair configurations begin with cutting out the sides as shown on the layout drawing. They can be used as-is for the basic chair. If you've decided to build the Rok-Bak, we'll cut the V-shaped bottom edges and bottom cutouts later.

It's a good idea to use a circular saw on the straight cuts, for speed, accuracy, and to minimize sanding. Notice that the gap between plywood pieces on the layout is at least 1/8". Thus one cut serves two pieces.

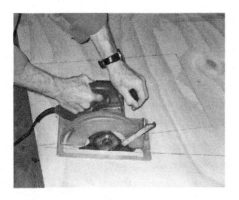

Note

You can minimize jigsaw splintering by using a splinter guard that snaps into the saw's footplate. Also, use new blades designed specifically to cut plywood. And if your jigsaw has multiple modes, set it to straight up-and-down rather than orbit mode. The cutting will go slower, but you'll get far fewer splinters.

After cutting one each of the chair and stool sides, sand all the edges smooth. Use a sanding drum accessory in your drill if you have one; it

will ensure nice, accurate, rounded corners that blend smoothly into the straight edges.

Now trace around the finished sides onto the plywood sheet, as shown in the layout drawing and photos. Once more: you must lay out mirror images so that the good side of the plywood will face out on the finished chair. Cut out the remaining sides.

Lay the chair sides face to face with the bad surfaces out, and drill the three 1/4" holes for attaching the sides to the cross braces. To minimize splintering, place scraps of wood under the side where the bit will exit. Repeat this process for the mounting holes in the stool sides.

Cut the chair's seat and back 22" wide to accommodate standard-sized patio chair cushions. For narrower cushions, adjust the widths accordingly. Add three countersunk holes to the back, following the layout.

Cut out the footstool top and the headrest support, backstop, and disks.

Step 2: Make the Cross Braces

Next cut three 22" lengths of 2 × 4 for the chair. These will be cross braces A, B, and C. You'll modify A and B to accommodate the headrest and backstop. (Again, if your cushions are less than 22" wide, adjust the length of the 2 × 4s to match the width of the seat and back.)

After cutting, draw an *X* on all the ends of the 2 × 4s, corner-to-corner. At the center of each *X*, drill a 5/32" hole at least 2" deep.

Cross brace A: If you plan to make a headrest, you need to shape a mounting surface for the headrest support piece. Using a hand or power plane, create a smooth, flat 1-1/2"-wide surface, at least 10" long, at approximately the angles shown. (Easier would be to cut it on an adjustable-table band saw, for the entire 22" length of the brace.) The 53° and 37° angles ensure that the headrest support is vertical before the chair is rocked back.

Cross brace B: Attach the backstop, a 22" × 1-1/4" × 1/2" strip of plywood that will support both seat and back pieces. This extra piece of plywood catches the bottom of the back, which

then stops the seat. Use three #8 × 1-1/2" wood screws, spaced approximately 8" apart.

Cross brace D: For the footstool, cut a 15" length of 2×4 and drill holes in the ends as you did on the other braces. We'll make the other miscellaneous parts after test-assembling the chair.

Step 3: Assemble the Basic Chair

Prop the sides apart while you fasten the cross braces between them using the lag bolts and washers.

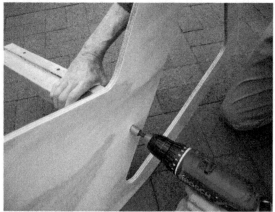

 Warning!

Don't scrimp on the size of these bolts, or your chair and/or stool could disassemble itself dramatically! Before you add the seat and back pieces to the chair, check for smoothness of all surfaces and edges. This is not only an appearance issue, but an obvious comfort issue as well.

Drop the back (three screw holes up) and seat (any orientation) into their respective positions. Notice that the bottom of the back fits between the rear edge of the seat and the backstop on cross brace B. Do not sit yet!

IMPORTANT: With the seat and back pieces flush with the wide side of all three cross braces, fasten the back to cross brace A with three #8 × 1-1/2" flathead wood screws. These screws prevent braces A and B from rotating; your weight will keep brace C fixed.

Now throw in a patio chair cushion set and give your chair a test-sit. The back cushion should rest on top of the seat cushion, whether they're a stitched assembly or separate cushions. I'm 5'11", 150 lbs, and the chair fits me perfectly. If you're shorter, you can trim the sides along the bottom so that your feet will be well supported when you sit in the chair.

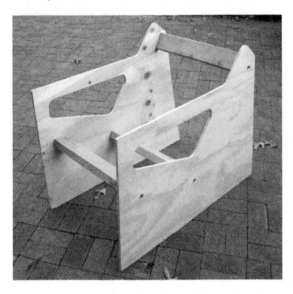

If you're sure you want to build only the basic configuration, you can make the large bottom cutouts in the sides as shown in the plans at *http://bit.ly/rokbakgithub*. However, if you're building the Rok-Bak version, leave the sides solid for now.

 Caution

Don't be tempted to make cutouts in the sides under the seat. This will *almost certainly make the chair too weak and wobbly.*

Step 4: Make and Attach the Arms

The only thing tricky about adding arms to the chair is the mounting itself. Each side of the chair will be taking 5 screws 5" apart, exactly centered in the top edge, which is 1/2" wide. Be careful: if these holes aren't drilled properly, the drill bit could break out the side.

You can control your drill bit better if you make a drill bit guide. It can be a piece of sheet metal, bent at a right angle or attached to a wood block. In any case, drill a 1" guide hole so that when using the guide, the guide hole will be exactly in the center of the plywood edge.

Draw lines on the top edge of the chair sides. Then using the drill-bit guide, drill five holes straight down into each side.

Make the arms from 1 × 4 fir or pine. If you plan to pad and cover the arms, this wood doesn't need to be the best quality, but if you'll leave it exposed, choose decent pieces of solid wood.

Now position an arm with the front of it overhanging the front of the chair side by 1/2". This overhang is necessary for stapling the fabric if you pad the arms. Mark the bottom surface of the arm to line up with the holes drilled in the top of the side. Drill the screw clearance holes from the bottom, which should ensure that the screw holes line up exactly. Repeat for the other arm.

Countersink the screw holes from the topside of the arms so the screws will be slightly sub-flush. Mount each arm with five #8 × 1-1/2" flathead wood screws.

Note

If you don't want to pad the arms, you could round the top-front corners of the arms with a 1/4"-radius router bit, or just sand something close. If you do want to pad the arms, highly recommended for comfort and appearance, that will be the very last step in building the chair (step 8).

Step 5: Make the Footstool Assembly

If you're making just the basic chair configuration, you can survive without a footstool. But for extra comfort, especially with the Rok-Bak, you'll need one.

Cut the parts and assemble the sides and cross brace D. You can make the top as 1 piece (see layout) or piece it together from cutout material (as shown here). Assemble the top and the 2 × 2 runners.

You can throw a pillow on the stool and help it stay in position with Velcro™, but covered padding looks more professional. Either way, the pillow or covering fabric should match your chair's cushions (if they're striped, you might choose a solid that's one of the stripe colors).

If you're padding, cut 1-1/2"-thick foam to match the top exactly. A band saw or electric knife works well. Cut the fabric about 4" oversize on all sides, pull it reasonably tight (the foam should compress slightly), and staple it in place.

Miter the corners. Use a minimum of staples at first, then add more as necessary to keep the fabric evenly taut, the foam reasonably compressed, and the corners neat.

Trim as shown.

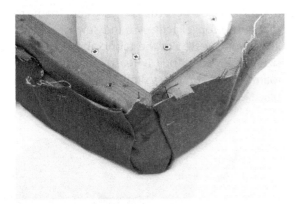

If you're sewing, sew an 8" Velcro™ strip to the visible side of the fabric, about 1" from one edge, centered. Sew all the way around the Velcro™ strip adjacent to the seam (on the outside of the sleeve), with one or more strips of bonding tape between the Velcro™ and the fabric. Iron the fabric side, not the Velcro™ side. This means you must put the iron in the sleeve.

Step 6: Make the Headrest

A well-padded headrest is key to a comfortable chair. Cover it with the same fabric you used for the footstool. For attaching the finished headrest to its support, you'll sew on Velcro™ at least 1" wide. Adhesive Velcro™ doesn't stick well enough to fabric.

Cut the fabric to 22" × 22". Secure the seam and Velcro™ strip. You have a choice of sewing or using fabric bonding tape. Use a sewing machine if you can. However, I found that Stitch Witchery tape makes a surprisingly strong seam, at least on denim.

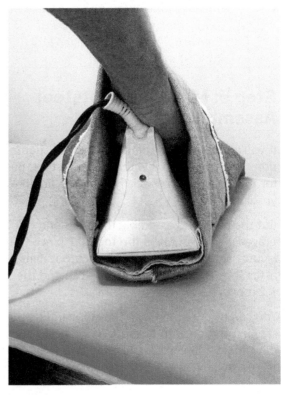

Cut a piece of 1-1/2" foam 19" × 12". Tightly roll up the foam from the short end, insert it into the fabric sleeve, and let it unroll. About 5" of the sleeve (22" − 12" = 10"/2 = 5") will overhang each end. The two ends of the foam should butt together inside and be "circularizing" the sleeve.

On one end, poke the surplus fabric over the foam and into the center hole. Cut the wood dowel, and drill both ends as shown. Attach one 4-1/4" disk to the end of the dowel with a 2-1/2" lag bolt and washer.

Press the dowel-disk assembly, dowel first, into one end of the cylinder. Gather the material as you press, and space the folds neatly around the inside circle. Stop pressing when the outside face of the disk is about 1-1/2" from the end.

From the other end, stuff at least 4 oz of Holofil or other stuffing between the dowel and the inside wall of the cylinder. Hold the other disk to keep it from being pushed out the end. Try to fill all voids, keeping the dowel in the center. The more stuffing you can push in, the more comfortable your headrest will be.

Fold the surplus material over the foam at the other end.

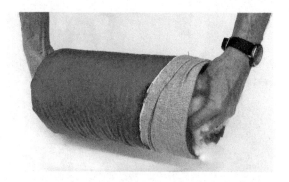

Put a lag bolt and washer through the hole in the second disk, then push the tip of the bolt into the hole in the end of the dowel. Tip the disk inward and press it into the foam core, keeping the fabric as neat as you can. You can press the other end's disk in a little farther to make it easier.

Tighten the second lag bolt with a ratchet wrench, while holding the first bolt with pliers. Neaten your fabric folds at both ends.

Staple two small strips of Velcro™ to the headrest support near the top, so their outside edges are 8" apart.

Mount the support to the bevel on cross brace A, exactly in the center, with three #8 × 1-1/2" screws. Temporarily stick the headrest to the Velcro™ strips on the support and give the chair another test-sit.

Step 7: Convert to Rok-Bak Version (Optional)

To allow the chair to rock back, the bottoms of the sides must be cut into shallow *V*s. The tipping point of the *V* is critical. Your Rok-Bak chair must:

- Be completely stable when rocked backward; do not exceed the 5-1/2" rocker dimension on the drawing.
- Stop, with a small thump, in the reclined position.
- Not lean prematurely; it should require a light push with your feet but not be difficult to rock backward.

Temporarily disassemble the chair so that you can make identical, accurate cuts (preferably with a circular saw) on the bottom of each side. Make it a shallow *V*, and if you're about the same height and weight as me (see step 3), you can make the bottom cuts as dimensioned here.

If you're of a different build, you should make a few trial cuts and reassemblies, taking longer and longer cuts until you arrive at the right balance. IMPORTANT: You must not exceed the 5-1/2" rocker dimension for safety and stability! After each pair of cuts, reassemble the chair—with cushions, headrest, and arms—and try it. You'll get pretty good at loosening and tightening lag bolts. (Use a 7/16" hex driver in your drill to speed things up.)

 Note

All is not lost if you cut too much off; you can always take a little off the front legs of the Vs to make them longer.

Once you get the tipping point right, you can finally make the bottom cutouts in the sides.

This will cause the chair to be slightly more stable in the forward position, which may turn out to be a good thing.

Draw the cutouts (on the C side, remember?) of both chair sides, using the 15oz can as a radius guide. For strength, it's important to keep the bottom of the cutouts at least 2-3/4" from the bottom edge of the chair. Carefully make the cutouts with a jigsaw.

Step 8: Finish the Wood and Pad the Arms

Remove the cushions, headrest, seat, and back. You don't have to disassemble the chair. Sand all exposed surfaces down to about 120-grit paper, then spray (or brush, if you must) with varnish or clear polyurethane, such as Deft. Several coats, lightly sanded between each, usually yield a nice finish.

Finally, to pad the arms, cut 3/4"-thick foam to the size of the arms (you can also use 1", or you can rip 1-1/2" in half with a band saw), and lightly mount it with spray adhesive or double-sided mounting tape. Cover the arms with fabric that matches the footrest and headrest. Trim the underside neatly.

Step 9: Make Yourself Comfortable

Now drop in your cushions, stick the headrest on, and pull up the footstool. Take a seat and make yourself comfortable. As you would with any rocking chair, watch out for the cat's tail, then rock back and dream of your next project: adding speakers to the headrest.

Like origami, single-sheet plywood projects transform a standard plane into countless 3D objects. Generations of designers have worked within this form, laying out cleverly fitted pieces that make furniture and toys with little or no wasted wood.

The Lost Art of Lashing: A Photo Essay

By Gever Tulley

Figure 8-1 *Our civilization was built on a technology so advanced, we still don't kno[w]
everything it's good for.*

re 8-2 *But somewhere along the way, most of us have forgotten how to tie things together.*

Figure 8-3 *Yet, if you can tie things together securely, you can make almost anythi*
from practically nothing.

Figure 8-4 *Inside a golf ball, a long rubber band subjects the core to 10,000 pounds of pressure. The secret is in the wrapping; every turn adds pressure.*

Figure 8-5 *Which is the same principle at work in a lashing. The result is very strong, but with a little bit of give.*

Materials List

- ☐ Twine (6lb test minimum)
- ☐ knife (for cutting twine and carving notches)
- ☐ Short Stick *

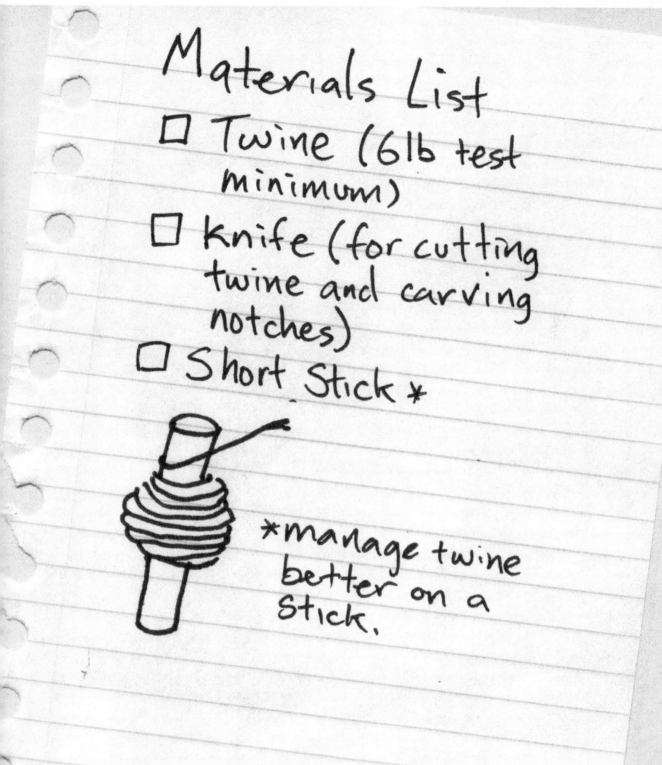

*manage twine better on a stick.

Figure 8-6 *OK, get your sticks and line, and we'll tie our first lashing.*

Figure 8-7 *This takes a little bit of practice, so expect to start over a couple of time*

Figure 8-8 *Always keep the line tight and tidy as you lay it down. Keep going until you have enough wraps.*

Figure 8-9 *8 to 10 wraps for twine and string.*

Figure 8-10 *Then, to really crank up the tension, add crossing wraps.*

Figure 8-11 *Then tie it off with a surgeon's knot (which is just an overhand with an extra twist.) Almost done.*

gure 8-12 *Now we lock the surgeon's knot in with a square knot. There are fancier ways to do this, but this is sufficient and easy to remember.*

Figure 8-13 *It is critical that the lashing is tight.*

ure 8-14 *So if one joint gets loose, you can whittle a wedge and pound it under the line to tighten it.*

Figure 8-15 *If you need to make a longer stick from two short ones, overlap and la* '*em.*

Figure 8-16 *Add a few loops across the lashing to tighten it up nicely.*

Figure 8-17 *Tying three sticks can be tricky, so instead of lashing all three together*

Figure 8-18 ... *you can just lash them in pairs.*

Figure 8-19 *You can run the lashing between many sticks at once to create a quick decking.*

Figure 8-20 *A secure, flexible connection is made by running the lashing between the sticks.*

Figure 8-21 *The same technique can accommodate three sticks for a tripod. These* *tripods are stable and strong and can be a foundation for other structures.*

Figure 8-22 *Some durable line is used to keep the legs from splaying.*

Figure 8-23 *A natural fork can add strength to your construction. And you only ha[ve]*
to lash to the strongest side; gravity does the rest.

Figure 8-24 *If the line slips on the wood, carve out a couple of notches to give it some purchase.*

Figure 8-25 *Making sketches always helps. And when you need real compression, few long loops around a pair of poles...*

Figure 8-26 *... and a short stick make for a highly effective turnbuckle.*

Figure 8-27 *Not everything has to be sticks. We'll just finish it off with a heavier li[?] here to make a railing.*

Figure 8-28 *Stable and strong; you can make almost anything with lashing!*

PART III
Evening

Pickle Grapes and Beets at Home

8

—Quick and tasty, vinegar pickling is great for preserving a surprising variety of fruits and veggies.

By Kelly McVicker

When you hear the word pickle, what comes to mind? A salty brined cucumber, slices of mango preserved in oil, or maybe a bowl of pungent kimchi? Pickling is a tradition found in virtually every culture on Earth. Mango pickles from India taste nothing like German sauerkraut, but the underlying process is the same.

Pickling transforms flavor and preserves food by raising the acidity to prohibit the growth of microorganisms that cause spoilage. The acid can be added in two ways:

Fermenting—This involves adding salt and allowing the food to sit for a period of time, while beneficial bacteria transform natural sugars into lactic acid.

Vinegar pickling—In this process, the acid comes from vinegar. Food is packed in a vinegar-based brine and allowed to sit for anywhere from a few hours to several months to achieve full flavor.

As a kid growing up in Kansas, I learned vinegar pickling from my grandmas and my mom. Even though I've become increasingly enchanted with fermentation, I still turn to the vinegar method when I want to make several jars of the same thing and store them without refrigeration, whether for holiday gifts or just for stocking for my own pickle pantry.

Tools

- Mason jars, pint or quart, with two-piece lids. I recommend using button-top lids.

- Medium saucepan.

- Measuring spoons.

- For Canning Beets
  — Water bath canner or stockpot with metal rack large enough to cover jars with 2" of water
  — Canning tongs (optional).
  — Wide-mouth funnel (optional).
  — Lid lifter (optional).

Spiced Pickled Grapes

Step 1: Assemble the Ingredients

Vinegar pickling usually gives more consistent results, since it quickly "shocks" the fruit or veggies to preserve them rather than waiting on the creation of new bacteria. And while vinegar pickling doesn't create probiotics, it does keep most of the nutrients intact. It's also more suitable for preserving foods without refrigeration. If you seal your pickles using the water bath canning process described in the recipe for Gingery Golden Beets, you can store them at room temperature for up to a year.

I like these two recipes because they show just how far beyond cucumbers your pickling experiments can take you. The grapes are sharp and a bit spicy; the beets mildly sweet with a ginger-plus-vinegar tang. Both pair deliciously with creamy cheeses. I like to toss them both with some arugula, goat cheese, and a little vinaigrette made from the brine for a pickled salad snack.

Grapes are perfect for pickling: firm and naturally acidic, with a soft skin that brine can permeate quickly. This recipe gives a deep, sweet flavor with surprising hints of savory from the garlic and bay leaf. Use Sichuan peppercorns if you can, for that special tongue tingle:

- 3 lbs red seedless grapes, such as Flame Seedless or Ruby Roman, firm but ripe

- 1-1/2 cups apple cider vinegar

- 1-1/2 cups red wine vinegar

- 1-1/2 cups water

- 1-1/2 cups sugar

- 2 tsp salt

- 3 tsp Sichuan peppercorns or black peppercorns

- 3 tsp mustard seeds
- 2" piece of ginger, sliced thinly
- 2 garlic cloves, sliced thinly
- 3 segments star anise
- 3 bay leaves

Step 2: Preparation

Remove all stems from the grapes. Rinse, pat dry, and set aside.

In a medium saucepan, combine vinegars, water, sugar, and salt and bring to a boil, stirring occasionally to dissolve. Reduce to a low simmer.

Distribute the spices evenly between the jars, except for a few pieces of ginger and garlic, and the bay leaves.

Step 3: Packing

Fill jars with grapes, taking care not to crush them, and leaving at least 1" of space at the top. Add the last pieces of garlic, ginger, and bay leaves around the outside of the jars.

Step 4: Pouring

Pour hot brine over the grapes, submerging them all.

Step 5: Storing

Put lids on jars, then let cool to room temperature before refrigerating.

Let sit in the refrigerator 48 hours before tasting. These grapes will keep in the refrigerator for up to two months, their flavor intensifying with time.

Gingery Golden Beets

Step 1: Assemble the Ingredients

Golden beets are milder than red ones. Because beets are dense and naturally low in acid, you'll take a few extra steps for pickling them. The water bath canning process creates a vacuum seal that makes the jars shelf-stable for a year.

- 4 lbs small golden beets

- 3-1/2 cups white vinegar
- 1-1/2 cups water
- 1 cup sugar
- 2" piece of ginger, sliced very thinly
- 20 black peppercorns

Step 2: Preparation

Bring a large pot of water to a boil. If you'll be canning your pickles, also bring a canning pot or another large stockpot to boil.

Remove beet tops, then scrub clean. Cook beets in boiling water until tender enough to pierce with a fork, about 10–15 minutes. Drain and cover with cold water to stop the cooking.

Combine vinegar, water, and sugar in a medium saucepan and slowly bring to a boil.

Peel and cut beets into medium-sized cubes or 1/2"-thick rounds.

Step 3: Packing

Pack the beets into clean pint-sized mason jars, leaving 1" of headspace.

Pour the hot liquid over the beets, covering them completely but leaving 1/2" of headspace. Use a chopstick or other nonmetallic instrument to remove air bubbles, then add more brine if necessary to keep 1/2" of headspace.

Step 4: Sealing

Seal the jars. Wipe rim with a paper towel, center a clean flat canning lid on top, screw on the lid band, and tighten to resistance. Don't over-tighten, or air can not escape during canning. (If skipping the canning process, refrigerate once the jars cool.)

Step 5: Boiling

Using canning tongs or another heatproof utensil, put jars in a canner or a large stockpot with enough water to cover the tops by 1"–2". Boil 25 minutes.

Step 6: Testing

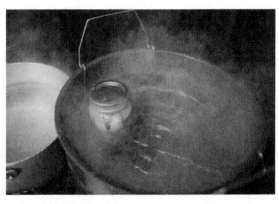

Turn off heat and remove jars to cool to room temperature. After an hour, check for a seal by pressing down on the indentation in the middle of the lid. (If the button can be pushed down, the jar hasn't sealed—store it in the fridge and consume within a month.)

Sealed jars will keep at room temperature for one year. After opening, refrigerate and consume within a month.

Kelly McVicker is the owner of McVicker Pickles. After spending several years raising money for women's human rights, she returned to her 4-H roots and jumped headfirst into the cutthroat world of small-batch pickling. She's a fiend for farmers markets and goes bonkers for a good brine.

Brine and Roast a Turkey

10

By Katie Goodman

Now that Halloween is over, I feel like the holiday season is ready to begin. I love all the cooking and baking that happens this time of year, but most of all, I love how the family gatherings always end up in the kitchen.

I'm lucky that as a young bride, my mother-in-law gave me some excellent tips when it was time for me to host my first Thanksgiving. I especially appreciated her tips on how to roast a turkey. She introduced me to brining, something I had never before heard of, but I knew that the turkey I had eaten at her home was the best I'd ever had, so I followed her advice. Here is my twist on the brine recipe she first gave me, as well some great tips for roasting a flavor-

ful turkey. Anyone who's ever eaten my turkey says it's the best they've ever had. And it's all thanks to my mother-in-law. I'm lucky to have such a sweet one!

The leftover turkey carcass from this recipe makes the best homemade turkey broth, which will go great alongside your side dishes and pumpkin pie.

Brine

Ingredients

- 16-pound turkey
- 3 c kosher salt
- 1 c brown sugar
- 1-1/2 tsp pepper
- 4 bay leaves
- 4 stems fresh thyme
- 3 stems fresh sage
- 1 Tbs minced garlic
- 1 gallon boiling water
- 8 pounds ice cubes

Step 1: Mixing

Stir together the salt, brown sugar, pepper, bay leaves, thyme, sage, and garlic in a large stock pot. Add 1 gallon of water. Bring to a boil, remove from heat, and allow the mixture to steep for 25 minutes. Stir in enough ice to bring the brine amount up to 2 gallons (2 gallons = 32 cups).

 Note

If your pot is not large enough, you may have to allow the brine to cool and add the additional amount when pouring the brine into the bag in the following step.

Step 2: Soaking

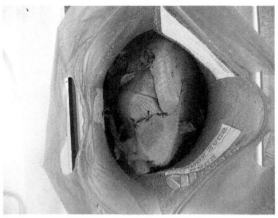

Place the turkey in a large zip-top bag. I recommend the Ziploc Big Bags (*http://www.ziploc.com/?p=b8*) (size large). Put the bagged turkey in a clean cooler. Pour the brine over the turkey, in the bag, making sure the breasts are fully submerged. Zip the bag closed. Place the cooler in a cool place, such as your garage, and allow the turkey to soak in the cold brine for 12-24 hours.

Use gel packs or bagged ice around the zipped bag inside the cooler, if necessary, to keep the brine below 40°F. (Adding more ice directly to the brine would only dilute it.)

After the brining process, transfer the turkey to a roasting pan and discard the brine.

 Note

For a smaller turkey, you may make less brine; however, be careful to do so with the original proportions of ingredient still intact. Too much salt will leave you with an incredibly salty turkey. Also, birds less than 10 pounds will likely not need to soak for the full 24 hours to achieve the desired results.

Garlic Herb Butter

Ingredients

- 8 Tbs butter, softened
- 3 cloves garlic, minced
- 3 tsp fresh thyme leaves
- 3 tsp fresh sage leaves, chopped
- 1 tsp fresh rosemary, minced
- 1/2 tsp black pepper

Step 1: Mixing

Combine all the ingredients in a bowl.

Step 2: Rubbing

Rub the butter all over the inside of the turkey cavity. Lift the skin covering the turkey breast meat and rub butter on top of the meat. Place the skin back down. Continue with roasting instructions.

Roasting

Step 1: Placing

Adjust the oven rack to the lowest position. Preheat the oven to 400°F. Place a V-rack in your roasting pan. Place the turkey, breast side down, on the V-rack. Pour two cups of water in the bottom of the pan.

Step 2: Stuffing

Place a quartered carrot, celery, and onion inside the turkey cavity if you aren't stuffing it. An unstuffed turkey will cook faster. If you are stuffing the turkey, do not tightly pack it in.

Step 3: Roasting

Tie the legs together using kitchen twine. Fold the wing tips under the turkey. Drizzle olive oil or melted butter on the outside of the turkey, if desired. Roast for 90 minutes.

Step 4: Checking

Remove the turkey from the oven and turn it breast side up. It helps to have a big wad of paper towels in each hand so you can easily flip it without slipping or burning yourself. Baste the turkey with pan drippings. Add an additional cup of water to the pan. Roast the turkey for an additional 40-65 minutes or until the meat thermometer inserted in the breast registers 160-165°F and the leg/thigh registers at about 170°F. Check to make sure the pan drippings and water have not completely evaporated, causing the herbs to burn. If the turkey begins to brown too much before it has reached the correct temperatures, cover it with foil.

Step 5: Resting

Transfer the turkey to a platter and allow it to rest for 20 minutes before carving. Use this time for any last minute preparations. This is the perfect time to put the rolls in the oven, make gravy or a salad, or set the table.

Roasting Pumpkin Seeds

By Katie Goodman

After you've made homemade pumpkin puree or carved your pumpkins for Halloween, don't throw the seeds away! They make a great snack when you roast them. Roasting pumpkin seeds is really simple, just watch and see.

Ingredients and Materials

- 1-1/2 cups pumpkin seeds
- 2 teaspoons butter, melted
- 2 teaspoons olive or canola oil
- Salt, to taste
- Cumin, to taste
- Paper towel
- Foil
- Oven
- Baking sheet
- Wooden spoon

Step 1: Rinse

93

Rinse all the excess pumpkin flesh and gunk off of the seeds.

Step 2: Place

Lay the seeds out on a paper towel to dry. (This can be done days in advance. If you're preparing your seeds earlier than you need, just place the completely dry seeds in an airtight container or resealable bag.)

Step 3: Heat

Meanwhile, line a baking sheet with foil and preheat the oven to 350 degrees F.

Step 4: Drizzle

Combine the melted butter and olive oil. Spread the seeds on the lined baking sheet. Drizzle with the oil and butter mixture and toss to coat.

Step 5: Redistribute

Redistribute the seeds so they are spaced out across the pan.

Step 6: Season

Season to taste with salt, and if desired, another spice (such as cumin or cinnamon).

Step 7: Roast

Roast at 350 degrees F for 20-30 minutes, stirring to toss the seeds halfway through.

Step 8: Eat

Remove and cool until you are able to handle the seeds. Eat and enjoy!

Pumpkin seeds are great eaten alone as a snack, but they also serve as a great garnish or topping for a variety of dishes, from desserts to salads.

PART IV

Night

Da Vinci Reciprocating Mechanism

12

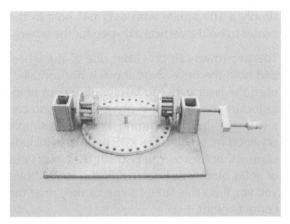

Figure 12-1 *Recreate Renaissance tech to convert rotation into a square wave.*

By Alan Federman

Last year I was blown away by the clever mechanisms displayed in the Leonardo da Vinci exhibit at the Tech Museum of Innovation (*http://thetech.org*) in San Jose. I was especially taken by da Vinci's simple mechanism for powering a sawmill with a water wheel. I made my own tabletop model of the mechanism, and it never fails to gather a crowd when I show it off.

The typical mechanism for converting between rotation and reciprocal motion is a flywheel and a crankshaft, like on a steam locomotive. But da Vinci's simple device starts with two modified cage gears (aka lantern gears) that ro-

tate on a common shaft. Each cage has half of its teeth missing, on opposite sides of the shaft from the other gear, so that when turned, they alternately engage with the pegs on opposite sides of a large wheel.

When you turn the shaft at a constant rate, this mechanism generates square-wave motion, rather than the sinusoidal motion of a piston. This is because each cage gear turns the wheel at a constant speed, and it lets go right before the other one comes pushing full-speed to turn the wheel the other direction.

Materials and Tools

- 1/4" birch plywood, 12" × 18"
- 1/4" wood dowel, 36" long
- 1/2" × 1/2" square wood molding, at least 4" long aka square dowel
- Wood glue
- Band saw or jigsaw
- Drill press with circle cutter and 1/4", 17/64", and 9/32" bits
- Laser cutter (optional) instead of the band/jigsaw and drill press
- Small saw for cutting dowels
- File, sanding block with medium-grit sandpaper, or Dremel with sanding bit

- Glue gun with hot glue C-clamps (2)
- Spray adhesive and computer with printer for the paper templates, if you don't use a laser cutter

Step 1: Cut the Pieces

I built my initial prototype by hand, and it took a lot of trial and error to get the peg spacing right. So I made a more precise second version by laying out the plywood pieces in Google SketchUp then importing them into CorelDraw and using the Corel files to cut them on an Epilog laser cutter.

The main plywood pieces you need to cut are four identical discs for the cage gears and a larger disc for the wheel (Figure 12-2). Each cage gear disc is 2.06" in diameter, with 10 equidistant 1/4" holes spaced 0.84" away from its center point. The wheel is 6.4" wide with 36 equidistant 1/4" holes spaced 2.95" from its center.

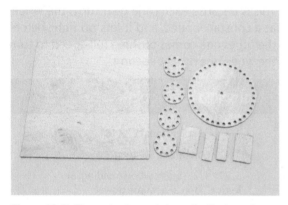

Figure 12-2 *The main plywood pieces for the base, large gear, two cage gears (aka lantern gears), and one of the two towers.*

To match da Vinci's plan, I also cut eight interlocking rectangular pieces that assemble into the two towers that hold the crankshaft. You can download my SketchUp and Corel files at http://bit.ly/davincigithub, along with full-size PDF templates that you can use to make the pieces by hand.

If you're cutting by hand, print out the three full-size templates, cut the shapes out of the paper, and temporarily affix them to the plywood with spray adhesive. Drill all the holes before you cut the discs out of the plywood, 17/64" holes in the center and 1/4" holes around the perimeter. Functionally, the mechanism needs only 26 holes cut, 4 in each cage gear disc and 10 in the large wheel. But I cut holes all the way around the perimeters of all the discs, 76 total, so that I could use the parts for other purposes, such as a crank assembly for a crane.

In addition to the discs, you need plywood pieces for the base and the towers. The base is simply a 10" square with a 17/64" hole in the center to hold a vertical axle peg for the wheel.

The two towers sit on either side of the wheel and hold the drive shaft above it horizontally. I used the laser cutter to fashion them out of interlocking pieces of 1/4" plywood, and you can also make them by cutting and gluing simple plywood rectangles; the PDF templates include plans for both. You can also cut the towers out of solid blocks. But don't drill the tower holes yet; you'll do this later, to ensure they're at the correct height.

From the 1/4" dowel, you need to cut 1 12" length for the drive shaft, 1 2" length for the hand crank, 8 1-1/4" pegs for the cage gears, 1 1" peg for the wheel's axle, and 10 5/8" pegs for the wheel's gearing (Figure 12-3). If you mark the pegs for cutting all at once, include some extra length for the width of the saw blade, or else the pegs might wind up too short.

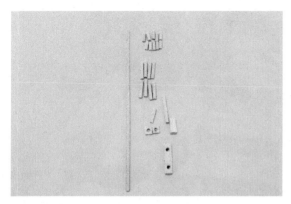

Figure 12-3 *Dowels, wooden washers, and crank assembly parts. The same dowel is used to make the crankshaft and gear pegs.*

From the 1/2" square dowel, cut and drill two thin slices to make washers that will fit over the drive shaft. Also cut a 2" block, and drill two parallel 17/64" holes through the block at opposite ends. Then cut another short piece, file it to make a round knob, and drill a blind 17/64" hole halfway into the knob.

Step 2: Assemble and Adjust

Build the two lantern gears by gluing four 1-1/4" pegs, positioned contiguously, between each pair of the smaller discs. Use wood glue and make sure the discs are aligned parallel (Figure 12-4).

Figure 12-4 *The two lantern gears after assembly.*

For the wheel axle, glue a 1" peg into the hole in the center of the base. Don't glue the towers together yet.

Sand or file one end of each 5/8" peg so that it's slightly rounded and tapered on two opposite edges; this helps the pegs engage with the cage gears (Figure 12-5).

Figure 12-5 *It's helpful to taper the pegs for the large gear; this will help them to mesh smoothly with the cage gears.*

Glue the pegs into the wheel, 5 in a row on opposite sides, with 13 empty holes (or undrilled hole positions) between them in each direction (Figure 12-6). The pegs should sit flush with the underside of the wheel, stick up 3/8" from the top face, and be oriented with their tapered edges facing the neighboring pegs. Let everything dry overnight. Get some sleep; tomorrow we have fun.

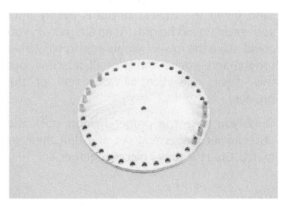

Figure 12-6 *The main gear assembled.*

Fitting the gears together is a bit tricky and takes some patience. First, fit the wheel over the axle.

Then mesh a lantern gear with the wheel's pegs on one side and use it to find the correct height for the crankshaft. This is the hardest part, but it's critical to smooth operation.

The best way is to drill two of the wide tower pieces a little above where you think the crankshaft should go. You can mark the point by running a ballpoint pen filler through the cage gear (Figure 12-7). Use a 9/32" bit so that the crankshaft spins freely.

Figure 12-7 *To measure the proper height for drilling the mounting towers, mesh a cage gear with the large gear's pegs, and mark where the shaft will meet the tower. Then drill slightly above your mark.*

Then repeatedly clamp down the two tower pieces, test the drive shaft through them, and sand down the tower's bottoms evenly until you reach a good height. (As an alternative, you could make the tower heights adjustable with a tongue-and-groove and a small setscrew, but this cheats the nice all-wood feel of the project.)

Once you have the proper height, stack the other tower pieces and cut and drill them to match. Glue the tower pieces together.

Figure 12-8 *It's important to drill the tower holes squarely, so the crankshaft won't act squirrely.*

For final adjusting, hot-glue the towers to the base. Assemble the crankshaft through the towers, hot-gluing the cage gears in place.

Figure 12-9 *Mounting the first tower. If it's a bit too high, you can sand down the bottom. If it's too low, you can shim it up.*

Work the mechanism in both directions and watch closely to see where any pegs stick, then sand down the pegs and cage gear dowels as needed (Figure 12-10).

Figure 12-10 *Adjusting the height of the peg teeth and rounding them slightly with a file is an alternative way to fine-tune the operation.*

When everything works smoothly, mark the precise locations of the cage gears on the crank- shaft, the towers on the base, and the shaft where it exits the towers; you'll glue the wooden washers here to prevent the shaft from sliding back and forth. You can put the washers inboard or outboard of the towers; outboard makes it easier to test-glue and adjust the washers.

Using wood glue, permanently reglue the towers to the base (rethreading the crankshaft if necessary to add your washers), and reglue the cage gears in place. Hot-glue the washers in place, test the mechanism again, then reglue them permanently.

Finally, glue the crank pieces onto the longer protruding end of the crankshaft, and crank away (Figure 12-11).

Figure 12-11 *Once everything works properly, add the crank.*

Oil Lamp from the Cave Dwellers of Lascaux

13

Figure 13-1 *What your iPhone is to you, the oil lamp may well have been to the cave dweller.*

By William Gurstelle

In Africa, Europe, and China, field scientists have uncovered the fossilized remains of campfire-charred bones so old that they likely pre-date Homo sapiens. The archaeological evidence suggests that our humanoid ancestors began taming fire perhaps as long as 1 million years ago.

While these creatures most likely lacked the wherewithal to kindle fire, they did, it seems, have the mental capacity to capture naturally occurring fire, tend it, and preserve it for long periods. For ancient hominids, campfires were

important not only for warmth and cooking, but also for light.

About 15,000 to 30,000 years ago, in the Late Stone Age (or Upper Paleolithic), humans painted elaborate images deep within several caves in western Europe, the best-known being those of Lascaux in southwestern France. Narrow and deep, the caves are impenetrable to daylight, making it impossible for the artists to have painted without some sustained source of artificial light.

Experts postulate that these primitive Rembrandts placed a few lumps of animal fat on a stone with a small manmade depression, then lit the fat with a burning faggot from the always-tended campfire not far away. The evidence indicates that in order to produce the hundreds of artworks now considered some of the world's oldest, the painters must have manufactured some of the world's first lamps as well.

As human culture progressed, so did lamp construction. Lamps were made from shells, bone, stone, and chalk, and were fueled by whatever naturally burning, organic substance was locally available. In the far north, it was whale blubber. In parts of the Middle East, lamps were fueled by petroleum, products such as liquid asphalt and naphtha, collected from seeps in the ground.

Today the ancient lamps most frequently depicted are those formed from fired clay that burned olive oil. African and Levantine lamps had open tops and were often hung on chains from the ceiling.

Later, great numbers of Roman lamps were manufactured using molds instead of hand-forming techniques. They're among the earliest examples of mass-produced housewares. Roman lamps had covers and sometimes multiple spouts and wicks that provided considerable light. It was in the orange-red glow of burning oil lamps that people like Aristophanes wrote, Socrates philosophized, and Archimedes invented.

Designing and fabricating a simple olive oil lamp is easy and fun, and quite possibly, useful. But best of all, when you do it, you form a connection with the technology of the past, of the earliest times of human civilization. What your iPhone is to you, the oil lamp may well have been to the cave dweller.

Materials and Tools

- 1 lb air-dry clay. Not all air-dry clays become waterproof when cured. For non-waterproof clays, the lamp interior may be coated with varnish or sealant if necessary, to prevent oil leaks.

- Fabric, 100% cotton, 3/4 inch wide, up to 4 inches long. Enough to make a wick.

- Olive oil. Enough to fill the lamp.

- Varnish, or glaze, waterproof if using non-waterproof clays. Enough to glaze the finished lamp.

- Scissors

- Scribes for inserting the wick into the spout, and for adding decoration (optional)

Step 1: Make the Lamp

Most clay or terra cotta oil lamps work the same way. The fiber skein or string that holds the flame is called the wick. Hydrocarbon compounds in the oil are wicked to the flame through the fibers via a phenomenon called capillary action. The upward-drawing motion of capillary action results from surface tension, or the attraction of molecules to molecules of similar kind. The oil molecules follow one another up the wick where they burn, or put a bit more scientifically, oxidize in a high-temperature, self-sustaining exothermic reaction.

An oil lamp is basically a reservoir for oil with a support to hold the wick upright. In practical terms this becomes a clay pot with a spout for a wick and a separate hole for adding the oil. Making a lamp on a potter's wheel is a simple task, as you need only throw a simple bowl, then pinch the wet clay to form a spout for the wick. You can even make a decent lamp with just your hands.

Step 2: Shape the Lamp

Olive oil lamps are simple enough to make without a potter's wheel. Almost any shape can be used, as long as it holds oil without leaking or spilling and has a spout and a filling hole. Once your lamp is shaped to your liking, follow directions on the clay package to cure and harden it. Air-dry clay may be used if a kiln isn't available; just paint the interior with varnish to make it waterproof if it's not already.

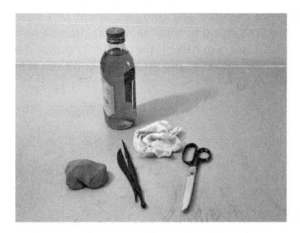

The simplest shape is a saucer lamp. Raised edges hold the oil, and a single depression in the rim forms the wick spout.

Over time, the saucer lamp was superseded by the covered lamp. The covered lamp has several advantages: it's less likely to spill, it usually has molded handles to make it easier and safer to transport, and the cover prevents contaminants from entering the oil reservoir.

Optional: Improve the finish of your lamp by lightly buffing it with cloth. Lamps can be detailed with scribes or knives, drilled, or sanded if desired. Lamps were commonly decorated with motifs that included mythology, animal and plant life, and repeating abstract designs.

Step 3: Make a Wick and Fill the Lamp

Cut a piece of cotton cloth 3/4" wide × 4" long (the exact length depends on the size of your lamp). Braid or twist the cloth in a tightly spiraled wick. Fill the lamp with olive oil.

Insert the wick into the lamp's spout. Using a scribe or other narrow tool, position the wick so it extends from the bottom of the oil lamp to approximately 1/2" above the spout. Trim the excess with scissors. Be sure the wick is saturated with oil.

Use under adult supervision only. Olive oil is flammable. Avoid spills, and use the lamp with care to avoid fire danger.

Olive oil produces a beautiful, soft orange flame but also considerable soot and smoke.

Carefully choose the location where you use the oil lamp to avoid getting soot on walls and ceilings. Oil lamps may set off smoke detectors.

Light the wick and enjoy the warm, soft light. You may need to trim the wick at intervals with the scissors to make it burn faster or slower depending on the amount of light you want it to produce.

Every issue of Make Magazine since Volume 4 has contained a piece written by William Gurstelle. He continues to develop new DIY stuff in his St. Paul, Minnesota workshop.

Index

Colophon

The cover images are photographs taken Ed Troxell, Julie Speigler, Samantha Gough, Tim King, and Helen Stewart. The cover fonts are URW Typewriter and Guardian Sans. The text font is Adobe Minion Pro; the heading font is Adobe Myriad Condensed; and the code font is Dalton Maag's Ubuntu Mono.

CPSIA information can be obtained at www.ICGtesting.com
Printed in the USA
BVOW10n0848021115

425145BV00001B/1/P